A Brief Account of Radioactivity

by Francis Preston Venable

I0484137

A BRIEF ACCOUNT OF RADIOACTIVITY

BY FRANCIS P. VENABLE, PH.D., D.SC., LL.D. PROFESSOR OF CHEMISTRY, UNIVERSITY OF NORTH CAROLINA AUTHOR OF "A SHORT HISTORY OF CHEMISTRY," "PERIODIC LAW," ETC.

D. C. HEATH & CO., PUBLISHERS BOSTON NEW YORK CHICAGO

PREFACE

I have gathered the material for this little book because I have found it a necessary filling out of the course for my class in general chemistry. Such a course dealing with the composition and structure of matter is left unfinished and in the air, as it were, unless the marvellous facts and deductions from the study of radio-activity are presented and discussed. The usual page or two given in the present text-books are too condensed in their treatment to afford any intelligent grasp of the subject, so I have put in book form the lectures which I have hitherto felt forced to give.

Perhaps the book may prove useful also to busy men in other branches of science who wish to know something of radio-activity and have scant leisure in which to read the larger treatises.

It is needless to say that there is nothing original in the book unless it be in part the grouping of facts and order of their treatment. I have made free use of the writings of Rutherford, Soddy, and J. J. Thomson, and would here express my debt to them--just a part of that indebtedness which we all feel to these masters. I wish also to acknowledge my obligations to Professor Bertram B. Boltwood for his helpful suggestions in connection with this work.

CONTENTS

CHAPTER I

DISCOVERY OF RADIO-ACTIVITY PAGE

CHAPTER II

PROPERTIES OF THE RADIATIONS

CHAPTER III

CHANGES IN RADIO-ACTIVE BODIES

CHAPTER IV

NATURE OF THE ALPHA PARTICLE

CHAPTER V

THE STRUCTURE OF THE ATOM

CHAPTER VI

RADIO-ACTIVITY AND CHEMICAL THEORY

A BRIEF ACCOUNT OF RADIO-ACTIVITY

CHAPTER I

DISCOVERY OF RADIO-ACTIVITY

The object of this brief treatise is to give a simple account of the development of our knowledge of radio-activity and its bearing on chemical and physical science. Mathematical processes will be omitted, as it is sufficient to give the assured results from calculations which are likely to be beyond the training of the reader. Experimental evidence will be given in detail wherever it is fundamental and necessary to a confident grasp of some of the marvelous deductions in this new branch of science. Theories cannot be avoided, but the facts remain while theories grow old and are discarded for others more in accord with the facts.

The Beginning

As so often happens in the history of science, the opening up of this new field with its fascinating disclosures was due to an investigation undertaken for another purpose but painstakingly carried out with a mind open to the truth wherever it might lead.

In 1895, Roentgen modestly announced his discovery of the X rays. This attracted immediate and intense interest. Among those who undertook to follow up these phenomena was Becquerel, who, because of the apparent connection with phosphorescence, tried the action of a number of phosphorescent substances upon the photographic plate, the most striking characteristic of the X rays being their effect upon such sensitive plates. In these experiments he obtained no results until he tried salts of uranium, recalling previous observations of his as to their phosphorescence. Distinct action was noted. Furthermore, he proved that this had no connection with the phenomenon of phosphorescence, as both uranic and uranous salts were active and the latter show no phosphorescence. Becquerel announced his discoveries in 1896 and this was the beginning of the new science of radio-activity.

Radio-active bodies

The rays given off by uranium and its salts were found to differ from the X rays. They showed no appreciable variation in intensity, no previous exposure of the substance to light was necessary, and neither changes of temperature nor any other physical or chemical agency affected them.

At first uranium and its compounds were the only known source of these new radiations, but many other substances were examined and two years later thorium and its compounds were added to the list. In general the discharging action seemed about the same. Other elements and ordinary substances show a minute activity. Only potassium and rubidium have a greater activity than this, and theirs is only about one-thousandth that of uranium.

An Atomic Property

In the examination of uranium and thorium compounds it was found that the activity was determined by the uranium and thorium present; it was proportto the amount ofional these elements present and independent of the nature of the other elements composing the compound. The conclusion was, therefore, that the activity was an inherent property of the atoms of uranium and thorium, that is, an atomic property. This was a long step forward and introduced into science the conception of a new property of matter, or at least of certain forms of matter.

Discovery of New Radio-active Bodies

In examining a large number of minerals containing uranium and thorium, Mme. Curie made the important observation that many of these were more active than the elements themselves. In measuring the activity she made use of the electrical method which will be described later. In the following table giving her results for uranium minerals the numbers under i give the maximum current in amperes. They serve simply for comparison.

i Pitchblende from Joachimsthal 7.0 $?10^{-11}$ Clevite 1.4 $?10^{-11}$ Chalcolite 5.2 $?10^{-11}$ Autunite 2.7 $?10^{-11}$ Carnotite 6.2 $?10^{-11}$ Uranium 2.3 $?10^{-11}$ Uranium and potassium sulphate 0.7 $?10^{-11}$ Uranium and copper phosphate 0.9 $?10^{-11}$

The last three are pure uranium and compounds of that element given for comparison with the first five, which are naturally occurring minerals. The last compound has the same composition as chalcolite and is simply the artificially prepared mineral. It has the activity which would be calculated from the proportion of uranium present, the copper and phosphoric acid contributing no activity.

Since the activity is not dependent upon the composition but upon the amount of uranium present, the activity in all of the minerals should be less than that of uranium. On the contrary, it is several times greater. Natural and artificial chalcolite also show a marked difference in favor of the former. The supposition was a natural one, therefore, that these minerals contained small quantities of an element, or elements, undetected by ordinary analysis and having a much greater activity than uranium. Similar results were obtained in the examination of thorium minerals and thorium salts.

Discovery of Polonium

Following up this supposition, M. and Mme. Curie set themselves the task of separating this unknown substance. Starting with pitchblende, a systematic chemical examination was made. This is an exceedingly complex mineral, containing many elements. The processes were laborious and demanded much time and minute care. They need not be described here. It is sufficient to say that along with bismuth a very active substance was separated, to which Mme. Curie gave the name of polonium for Poland, her native land. Its complete isolation is very difficult and sufficient quantities of the pure substance have not been obtained to determine its atomic weight and other properties, but some of the lines of its spectrum have been determined. Chemically it is very closely analogous to bismuth.

Discovery of Radium

In a similar manner a barium precipitate was obtained from pitchblende which contained a highly active substance. The pure chloride of this body and barium can be prepared together and then separated by fractional crystallization. To the new body thus found the name of radium was given. It is similar in chemical properties to barium. Its atomic weight has been determined by several careful investigators and is accepted as 226. Its spectrum has been mapped and its general properties are known. It is a silvery white, oxidizable metal. In one ton of pitchblende about 0.2 gram of radium is present; this is about 5000 times greater than the amount of polonium present. The activity of the products was depended upon as the guide in these separations. The radium found is relatively enormously more active than the pitchblende or uranium.

Other Radio-active Bodies Found

In the above separations use was made of relationships to bismuth and barium. Similarly, by taking advantage of chemical relationship to the iron group of elements, another body was partially separated by Debierne, to which he gave the name actinium. Boltwood discovered in uranium minerals the presence of a body which he named ionium, and which is so similar to thorium that it cannot be separated from it. It, however, far exceeds thorium in activity.

The lead which is present in uranium and thorium minerals--apparently in fairly definite ratio to the amount of uranium and thorium--is found, on separation and purification, to possess radio-active properties. This activity is due to the presence of a very small proportion of an active constituent called radio-lead, which has chemical properties identical with those of ordinary lead. The bulk of the lead obtained from radio-active minerals differs in atomic weight from ordinary lead and appears also to be different according to whether its source is a thorium or a uranium mineral.

A large number of other radio-active substances have been separated and some of their properties determined, but these were found by different means and will be noted in their proper place. They number in all more than thirty. The sources or parents of these are the original uranium or thorium, and the products form regular series with distinctive properties for each member.

CHAPTER II

PROPERTIES OF THE RADIATIONS

The activity of these radio-active bodies consists in the emission of certain radiations which may be separated into rays and studied through the phenomena which they cause.

Ionization of Gases

One of these phenomena is the power of forming ions or carriers of electricity by the passage of the rays through a gas, thus ionizing the gas. The details of an experiment will serve to make the meaning of this ionization clear.

When this apparatus is set up a minute current will be observed without the introduction of any radio-active matter. This, as Rutherford says, has been found due mainly to a slight natural radio-activity of the matter composing the plates. If radio-active matter is spread on plate A, which is connected with one pole of a grounded battery, and if plate B is connected with an electrometer which is also connected with the earth, a current is caused which increases rapidly with the difference of potential between the plates, then more slowly until a value is reached that changes only slightly with a larger increase in the voltage.

According to the theory of ionization, the radiation produces ions at a constant rate. The ions carrying a positive charge are attracted to plate B, while those negatively charged are attracted to plate A, thus causing a current. These ions will recombine and neutralize their charges if the opportunity is given. The number, therefore, increases to a point at which the ions produced balance the number recombining.

When an electric field is produced between the plates, the velocity of the ions between the plates is increased in proportion to the strength of the electric field. In a weak field the ions travel so slowly that most of them recombine on the way and consequently the observed current is very small. On increasing the voltage the speed of the ions is increased, fewer recombine, the current increases, and, when the condition for recombination is practically removed, it will have a maximum value. This maximum current is called the saturation current and the value of the potential difference required to give this maximum current is called the saturation P.D. or saturation voltage.

The picture, then, is this. The radiations separate the components of the gas into ions, or carriers of electricity, half of which are charged negatively and half positively. In the electric field those negatively charged seek the positive plate and those positively charged seek the negative plate. If time is given, these ions meet and recombine, their charges are neutralized, and there is no current.

Experimental Confirmation

This theory of the ionization of gases has been most interestingly confirmed by direct experiment. For instance, the ions may form nuclei for the condensation of water, and in this way the existence of the separate ions in the gas may be shown and the number present actually counted.

When air saturated with water vapor is allowed to expand suddenly, the water present forms a mist of small globules. There are always small dust particles in air and around these as nuclei the drops are formed. These drops will settle and thus by repeated small expansions all dust nuclei may be removed and no mist or cloud will be formed by further expansions.

If now the radiation from a radio-active body be introduced into the condensation vessel, a new cloud is produced in which the water drops are finer and more numerous according to the intensity of the rays. On passing a strong beam of light through the condensation chamber, the drops can readily be seen. These drops form on the ions produced by the radiation.

Application of Electric Field

If the condensation chamber has two parallel plates for the application of an electric field like that already described, the ions will be carried at once to the electrodes and disappear. The rapidity of this action depends upon the strength of the electric field and experiment shows that the stronger the field the smaller the number of condensation drops formed. If there is no electric field, a cloud can be produced some time after the shutting off of the source of radiation, showing that time is required for the recombination of the ions.

Size and Nature of Ions

If the drops are counted (there being special methods for this) and the total current carried accurately measured, then the charge carried by each ion may be calculated. This has been determined. The mass of an ion compared with the mass of the molecules of gas in which it was produced can also be approximately estimated. In the study of these ions the view has been held that the charged ion attracted to itself a cluster of molecules which surrounded the charged nucleus and traveled with it. It is roughly estimated that about thirty molecules of the gas cluster around each charged ion.

Photographing the Track of the Ray

Utilizing the fact that these ions with their clusters of molecules form nuclei for the condensation of water vapor, C. T. R. Wilson has by instantaneous photography been able to photograph the track of an ionizing ray through air. The number of the ions produced, and hence the number of drops, is so great that the trail is shown as a continuous line. In the copy of this photograph it will be seen that at some distance from its source the straight trail is slightly but abruptly bent. Near the end of its course there is another abrupt and much sharper bend. These bends show where the ionizing ray, in this case an alpha particle, has been deflected by more or less direct collision with an atom. These collisions and the final disappearance of the ray will be discussed later.

Action of Radiations on Photographic Plates

Taking up now other means of examining these radiations, it is well to consider their action upon a photographic or sensitive plate. It will be recalled that this was the method by which their existence was originally detected. To illustrate the method, the following account of how one such photograph was taken may be given.

The plate was wrapped in two thicknesses of black paper. The objects were placed upon this and the radio-active ore, separated by a board one inch thick, was placed above. The exposure lasted five days. The action is much less rapid and the result not so clearly defined as in the case of photographs taken by X rays. Of course, the removal of the board and the use of more concentrated preparations of radium would give quicker and better results. The method, however, on account of time consumed and lack of definition is ill adapted to accurate work.

Discharge of Electrified Bodies

The radiations from radio-active bodies can discharge both positively and negatively electrified bodies by making the air surrounding them a conductor of electricity. To demonstrate this, use is made of an electroscope. If the hinged leaf of such an instrument be electrically charged and a radio-active body be brought into its neighborhood, the electricity will be discharged and the leaf return to its original position. The rapidity of this discharge is used to measure the degree of activity of the body giving off the radiation.

The gold-leaf L is attached to a flat rod R and is insulated inside the vessel by a piece of amber S supported from the rod P. The system is charged by a bent rod CC' passing through an ebonite stopper. After charging, it is removed from contact with the gold-leaf system. The rods P and C and the cylinder are then connected with the earth.]

Scintillations on Phosphorescent Bodies

It was found by Crookes that a screen covered with phosphorescent zinc sulphide was brightly lighted up when exposed to the radiations. This is due to the bombardment of the zinc sulphide by a type of ray called the alpha ray. Under a magnifying glass this light is seen to be made up of a number of scintillating points of light and is not continuous, each scintillation being of very short duration. By proper subdivision of the field under the lens, the number of scintillations can be counted with close accuracy.

A simple form of apparatus called the spinthariscope has been devised to show these scintillations. A zinc sulphide screen is fixed in one end of a small tube and a plate carrying a trace of radium is placed very close to it. The scintillations can be observed through an adjustable lens at the other end of the tube. Outer light should be cut off, as in a dark room. The screen then appears to be covered with brilliant flashes of light. Other phosphorescent substances, such as barium platino-cyanide, may be substituted for the zinc sulphide, but they do not answer so well.

Penetrating Power

By penetrating power is meant the power exhibited by the rays of passing through solids of different thicknesses and gases of various depths. This power varies with different radiations and with the nature of the solid or gas. For instance, a sheet of metallic foil may be used and the effect of aluminum will differ from that of gold and the different rays vary in penetrating power. In the case of gases air will differ from hydrogen, and it is noticed that certain rays disappear after penetrating a short distance, while others can penetrate further before being lost.

Magnetic Deflection

If the radiations are subjected to the action of a strong magnetic field, it is found that part of them are much deflected in the magnetic field and describe circular orbits, part are only slightly deflected and in the opposite direction from the first, and the remaining rays are entirely unaffected.

Three Types of Rays

By the use of these methods of investigation it is learned that the radiations consist of three types of rays. These have been named the alpha, beta, and gamma rays, respectively. Some radio-active bodies emit all three types, some two, and some only one. The distinguishing characteristic of these types of rays may be summed up as follows:

Alpha Rays

The alpha rays have a positive electrical charge and a comparatively low penetrating power. They are slightly deflected in strong magnetic and electric fields. They have a great ionizing power and a velocity about one-fifteenth that of light.

Beta Rays

The beta rays are negatively charged and have a greater penetrating power than the alpha rays. They show a strong deflection in magnetic and electric fields, have less ionizing power than the alpha rays, and a velocity of the same order as light.

Gamma Rays

The gamma rays are very penetrating and are not deflected in the magnetic or electric fields. They have the least ionizing power and a very great velocity.

The penetrating power of each type is complex and varies with the source, so the statements given are but generalizations. The alpha rays are projected particles which lose energy in penetrating matter. As to the power of ionizing gases, if that for the [alpha] rays is taken as 10,000, then the [beta] rays would be approximately 100 and the [gamma] rays 1.

Measurement of Radiations

The rays are examined and measured in several ways:

1. By their action on the sensitive photographic plates. The use of this method is laborious, consumes time, and for comparative measurements of intensity is uncertain as to effect.

2. By electrical methods, using electroscopes, quadrant electrometers, etc. These are the methods most used.

3. By exposure to magnetic and electric fields, noting extent and direction of deflection.

4. By their relative absorption by solids and gases.

5. By the scintillations on a zinc sulphide screen.

Identification of the Rays

The alpha rays have been identified as similar to the so-called canal rays. These were first observed in the study of the X rays. When an electrical discharge is passed through a vacuum tube with a cathode having holes in it, luminous streams pass through the holes toward the side away from the anode and the general direction of the stream. They travel in straight lines and render certain substances phosphorescent. These rays are slightly deflected by a magnetic field and in an opposite direction from that taken by the cathode rays in their deflection. The rays seem to be positive ions with masses never less than that of the hydrogen atom. Their source is uncertain, but they may be derived from the electrodes.

The beta rays are identical in type with the cathode rays and are negative electrons.

The gamma rays are analogous to the X rays and are of the order of light. They are in general considerably more penetrating than X rays. For example, the gamma rays sent out by 30 milligrams of radium can be detected by an electroscope after passing through 30 centimeters of iron, a much greater thickness than can be penetrated by the ordinary X rays.

CHAPTER III

CHANGES IN RADIO-ACTIVE BODIES

Is Radio-activity a Permanent Property?

Is this power of emitting radiations a permanent property or is it lost with the passage of time? The first investigations of the activity of uranium and thorium showed no loss of intensity at the end of several years, and radium also seemed to show no decrease in its enormous activity. Polonium, however, was found to lose most of its activity in a year, and later it appeared that some radio-active substances lost most of their activity in the course of a few minutes or hours.

Induced Activity

A phenomenon called induced or secondary radio-activity was also observed. Thus a metal plate or wire exposed to the action of thorium oxide for some hours became itself active. This induced activity was not permanent but decreased to half its value in about eleven hours and practically disappeared within a week. Similar phenomena were observed when radium was substituted for thorium.

Discovery of Uranium X

In 1900 Crookes precipitated a solution of an active uranium salt with ammonium carbonate. The precipitate was dissolved so far as possible in an excess of the reagent, leaving an insoluble residue. This residue was many hundred times more active, weight for weight, than the original salt, and the solution containing the salt was practically inactive. At the end of a year the uranium salt had regained its activity while the residue had become inactive.

Another method of obtaining the same result is to dissolve crystallized uranium nitrate in ether. Two layers of solution are formed, one ether and the other water coming from the water of crystallization. The aqueous layer is active, while the water layer is inactive. Similarly, by adding barium chloride solution to a solution of a salt of uranium and then precipitating the barium as sulphate, the activity is transferred to this precipitate. These experiments give proof of the formation and separation of a radio-active body by ordinary chemical operations.

So, too, in the case of thorium salts a substance can be obtained by means of ammonium hydroxide which is several thousand times more active than an equal weight of the original salt. After standing a month, the separated material has lost its activity and the thorium salt has regained it. Here, again, there is the formation, separation, and loss of a radio-active body.

Conclusions Drawn

Now, these are ordinary chemical processes for the separation of distinct chemical individuals. The results, therefore, lead naturally to the conclusions: (1) it would seem that uranium and thorium are themselves inactive and the activity is due to some other substance formed by these elements; (2) this active substance is produced by some transformation in those elements, for on standing the activity is regained. This latter conclusion is startling, for it indicates a change in the atom which, up to the time of this discovery, was deemed unchangeable under the influence of such physical and chemical changes as were known to us.

Search for New Radio-active Bodies

The search for new radio-active bodies and the study of their characteristics has been systematically and successfully carried on. The bodies obtained in the above experiments were named uranium X and thorium X, respectively. Further, it became clear from the investigation of uranium minerals that radium, polonium, actinium, and ionium originated from uranium. From thorium minerals a body was separated called mesothorium, which was analogous to radium. Both thorium and radium were found to give off a radio-active gas. The first lost half of its activity in less than one minute. The second was more stable and lost half of its activity in about four days. The name radium emanation was given to the latter and it was found chemically and physically to belong to the class of monatomic or noble gases, such as helium, argon, neon, etc., which had been discovered by Ramsay. In some cases the chemical action was determined and these new bodies were found analogous to well-known elements, as radium to barium, polonium to bismuth. The physical properties were investigated and, where possible, spectra were mapped and atomic weights determined.

It is clear, therefore, that these bodies are elemental in character and as such are made up of distinct, similar atoms, just as the commonly recognized elements are believed to be. In this way more than thirty new elements have been added to the list. These new elements are called radio-active elements, but it is an open question whether all atoms do not possess this property in greater or less degree. Certainly, it is possessed in varying degree by four of the old elements widely separated in the Periodic System, namely, uranium, thorium, rubidium, and potassium. The last two, while feebly active themselves, do not form any secondary radio-active substance so far as is known. Only two of the elements, then, can definitely be said to go through these transformations. It is just possible that radio-activity may be found to be a common property of all atoms and of all matter.

Methods of Investigation

It is important to know how these new bodies were discovered and distinguished from one another. Two properties are relied upon. One is the nature of the rays emitted and the other is the duration of the activity. Of course, knowledge of the physical and chemical properties is also of great importance whenever obtainable.

Nature of the Radiations

The nature of the radiation is a distinguishing characteristic, though similarity here does not prove identity of substances. Some emit [alpha] rays only, some emit [beta] rays, some emit two of the possible rays, as for instance, [beta] and [gamma], and some emit all three. The rays may also differ in the velocity with which they are emitted by different radio-active substances. Thus, in the case of one substance the [alpha] rays may have a slightly greater or less penetrating power than those emitted by some other substance, and this may be true also of the other rays.

Life Periods

The duration of the activity is called the life period. This is absolutely fixed for each body and furnishes the most important mode of differentiating among them. It measures the relative stability and is the time which must elapse before their activity is lost and they, changing into something else, entirely disappear. The measure usually adopted is the half-value period. Two hypotheses are made use of:

1. That there is a constant production of fresh radio-active matter by the radio-active body.

2. That the activity of the matter so formed decreases according to an exponential law with the time from the moment of its formation.

These hypotheses agree with the experimental results. The decrease and rise of activity, for example, of uranium and uranium X, and also of thorium and thorium X, have been measured, plotted, and the equations worked out.

Manifestly, a state of equilibrium will be reached when the rate of loss of activity of the matter already produced is balanced by the activity of the new matter produced. This equilibrium and the knowledge of the rate of decrease in general will have little value if this rate, like chemical changes, is subject to the influence of chemical and physical conditions. The rate of decrease has been found to be unaltered by any known chemical or physical agency. For instance, neither the highest temperatures applicable nor the cold of liquid air have any appreciable effect.

Equilibrium Series

In order to measure the disintegration of a radio-active body in units of time so that the rate may be comparable with that of other radio-active bodies, the relation between the amounts under consideration must be a definite one. For this purpose equal weights of the bodies are not taken, but use is made of the amounts which are in equilibrium with a fixed amount of the parent substance.

One gram of radium has been settled upon as the standard for that series and a unit known as the "curie" has been adopted to express the equilibrium quantity of radium emanation. Thus, a curie of radium emanation (or niton) is the weight (or, as this is a gas, the volume at standard pressure and temperature) of the emanation in equilibrium with one gram of radium. This, by calculation and experiment, is found to be 0.63 cubic millimeter. When this amount has been produced by one gram of radium, the formation and decay will exactly balance one another. This is, therefore, one curie of emanation.

The measurement of the rate of decay is difficult but can be carried out with great accuracy, even down to seconds, in the case of certain short-lived bodies. Errors crept in at first from the failure to completely separate the substances produced in the series, and sometimes because of the simultaneous production of two substances.

As stated, the decay follows an exponential law. The time required for the decay of activity to half-value does not mean, therefore, that there will be total decay in twice that time. Thus the half-value period for uranium X is about 22 days. The period for complete decay is about 160 days. This half-value period corresponds to the half-value recovery period of uranium, which is also 22 days.

These were the earlier figures obtained for uranium X and they illustrate some of the difficulties surrounding such determinations. It was found later that the body examined as uranium X was really a constant mixture and of course the decay and recovery periods were also composite. It required later and very skilful work to separate them into the bodies indicated in the disintegration series.

The half-value period for thorium X is much shorter, namely, a little over four days, and this is also the recovery period for thorium X. The plotted decay and recovery curves will intersect at this point.

The consecutive disintegration series, with the half-value periods, for the uranium and thorium series as given by Soddy are seen in the following tables.

They are probably subject to some changes on further and more accurate determination. The nature of the rays emitted is also given.

Uranium (8 ?10^9 years) 238.5 -> [alpha] -> [alpha] \/ Uranium X (35.5 days) (230.5) -> [beta]&[gamma] -> ([beta]) \/

\/

\/

Ionium (5 ?10^4 to 10^6 years) (230.5) -> [alpha] \/ Radium (2,500 years) 226.4 -> [alpha] \/ Emanation (5.57 days) (222.4) -> [alpha] \/ Radium A (4.3 minutes) (218.4) -> [alpha] \/ Radium B (38.5 minutes) (214.4) -> ([beta]) \/ Radium C{1} (28.1 minutes) { (214.4) -> [alpha] { -> [beta]&[gamma] { \/ Radium C{2} (1.9 minutes) { (210.4) -> [beta]&[gamma] \/ Radium D (24 years?) (210.4) -> ([beta]) \/ Radium E (7.25 days) (210.4) -> [beta]&[gamma] \/ Radium F (Polonium 202 days) (210.4) -> [alpha] \/ Radium G (probably lead) (206.8)

Actinium (?) \/ Radio-Actinium (28.1 days) -> [alpha] -> ([beta]) \/ Actinium X (15 days) -> [alpha] \/ Emanation (5.6 seconds) -> [alpha] \/ Actinium A (0.0029 second) -> [alpha] \/ Actinium B (52.1 minutes) -> ([beta]) \/ Actinium C{1} (3.10 mins.) { -> [alpha] { \/ Actinium C{2} (?) { -> [alpha] \/ Actinium D (7.4 minutes) -> [beta]&[gamma] \/ Actinium E (unknown)

Thorium (4 ?10^{10} years?) 232.4 -> [alpha](?) \/ Mesothorium{1} (7.9 years) \/ Mesothorium{2} (8.9 hours) -> [beta]&[gamma] \/ Radiothorium (2.91 years?) -> [alpha] \/ Thorium X (5.35 days) -> [alpha] \/ Emanation (76 seconds) -> [alpha] \/ Thorium A (0.203 second) -> [alpha] \/ Thorium B (15.3 hours) -> ([beta]) \/ Thorium C{1} (79 minutes) { -> [alpha] { \/ Thorium C{2} (?) { -> [alpha] \/ Thorium D (4.5 minutes) -> [beta]&[gamma] \/ Thorium E (unknown)

FIG. 6.--DISINTEGRATION SERIES FOR URANIUM, ACTINIUM, AND THORIUM,

CHAPTER IV

NATURE OF THE ALPHA PARTICLE

Disintegration of the Elements

The remarkable disintegrations related in the last chapter, in which the heaviest known elementary atom--that of uranium (at. wt. 238)--is by successive stages changed into others of lower atomic weight, afford a clue to the nature of the atom and to that goal of the chemist, the final constitution of matter. The composite nature of the atom and some sort of interrelation of the elements had previously been made apparent from a study of the Periodic System and data gathered still earlier, but all attempts at working out a so-called genesis of the elements had proved vague and unsatisfactory.

Identification of the Rays

To get an understanding of the disintegration occurring in radio-active substances, the nature of the rays produced must be known. These rays are the cause of the activity and their emission accompanies the changes or disintegration. They have for the sake of convenience been called the alpha, beta, and gamma rays. The gamma rays have been identified with the X rays discovered by Roentgen and are a form of energy analogous to light. The beta rays are particles of negative electricity or electrons. With these, then, we have some degree of familiarity. But what are the alpha rays? An answer to this question should make clearer the character of the changes taking place, and should give some insight into the composition and mechanism of the atom.

The Alpha Rays

It has already been stated that these alpha rays are similar or analogous to the canal rays, but this advances the matter very little, as the nature of these canal rays has not been fully determined. The full identity with them, if proved, should have an important theoretical bearing.

Alpha Rays Consist of Solid Particles

In the first place, these alpha rays have been found to be made up of solid particles, that is, of what we are accustomed to call matter. Since it has become more and more difficult to draw a clear distinction between matter and energy, it would perhaps be better to say that these particles appear to have some of the properties hitherto attributed solely to matter. The best evidence that these particles are of atomic mass is furnished by their deflection in electric and magnetic fields.

Electrical Charge

It is not of first importance to discuss this or other proofs of the material nature of these particles. That they carry a charge of positive electricity is, however, a fact of very great import. The value of this charge has been carefully determined by a number of investigators working with different sources of the alpha particles and has been found to be 9.3 ?10^{-10} electrostatic units (.000,000,000,93 e.s.). From the consideration of the charge upon an electron previously obtained by J. J. Thomson and others, it was concluded that the alpha particle carried two unit positive charges; the fundamental unit charge, therefore, is half this value, or 4.65 ?10^{-10} e.s.

Helium Formed from Alpha Particles

To determine the nature of the alpha particle a crucial experiment was carried out by Rutherford and Royds, which was described as follows:

A large quantity of radium emanation was compressed into a fine glass tube A, about 1.5 cm. long. This tube, which was sealed to a larger capillary tube B, was sufficiently thin to allow the alpha particles from the emanation and its products to pass through, but sufficiently thick to withstand atmospheric pressure. The thickness of the glass wall was in most cases less than .01 mm. On introducing the emanation into the tube, the escape of the alpha particles from the emanation was clearly seen by the scintillations produced at some distance on a zinc sulphide screen. After this test the glass tube A was surrounded by a glass tube T and a small spectrum tube V attached to it. The tube T was exhausted to a charcoal vacuum. By means of the mercury column H, the gases in the tube T could at any time be compressed into the spectrum tube V and the nature of the gases which had been produced determined spectroscopically. It was found that two days after the introduction of the emanation into A the spectrum showed the yellow line of helium, and after six days the whole helium spectrum was observed. In order to be certain that the helium, coming possibly from some other source, had not diffused through the thin walls of the tube A, the emanation was pumped out and helium substituted. No trace of helium could be observed in the vacuum tube after several days, showing that the helium observed in the first experiment must have originated from the alpha particles which had been propelled through the thin glass tube into the outer tube.

Most of the alpha particles are propelled with such force that they penetrate some distance into the walls of the outer tube and some of these gradually diffuse out into the exhausted space. The presence of helium in the spectrum tube can be detected after a shorter interval if a thin cylinder of lead is placed over the emanation tube, since the particles fired into the lead diffuse out more rapidly than from glass.

A still more definite proof of the identity of the alpha particle with the helium atom was obtained by removing the outer glass tube T and placing a cylinder of lead over the emanation tube in the open air. Helium was always detected in the lead after it had remained several hours over the thin tube containing a large quantity of the emanation. In order to test for the presence of helium in the lead, the gases present were released by melting the lead in a closed vessel. There can thus be no doubt that the alpha particle becomes a helium atom when its positive charge is neutralized.

Thus the chemist was afforded the experience of the building up of at least one element under his observation, and both the analysis and synthesis of matter have been revealed through the discoveries of radio-activity.

Discovery of Helium

It is of interest at this point to learn something of the history of helium and its occurrence. In 1868 there was discovered by Janssen and Lockyer a bright yellow line in the spectrum of the sun's chromosphere. Because of its origin the name helium was given to the supposed new element causing it. Later it was found in the spectra of many of the stars, and because of its predominance in some of these they were called helium stars. Its existence on our planet was not detected for nearly thirty years.

In 1895, in connection with the discovery of argon in the atmosphere, a search was made to see if the latter element could be obtained from mineral sources. In analyzing certain uranium minerals Hillebrand had found considerable quantities of a gas which he took to be a peculiar form of nitrogen. Ramsay made a further examination of the gas coming from these minerals and the spectroscope revealed the yellow line of helium, thus at last proving the presence of this element on the earth. It is known now to be present in thorium minerals, in the waters of radio-active wells, and in minute amounts in the atmosphere. Its occurrence in every case, in the light of the experiment described above, would seem to be due to the presence of radio-active changes.

Characteristics of Helium

Helium, on account of its chemical inactivity and physical properties, is classed along with argon, neon, krypton, and xenon in the zero group of the Periodic System, and forms with them the monatomic, inert gases. In this class are now placed also the three radio-active gases, emanating respectively from radium, thorium, and actinium. These are generally known as radium emanation, thorium emanation, and actinium emanation. The first mentioned was once called niton. Emanium was the name originally proposed by Giesel for the body now known as actinium.

The calculated rate of production of helium in the series in equilibrium with one gram of radium is 158 cubic millimeters per year. This corresponds quite well with the experimental results.

Table of Constants

Some of the more important atomic and radio-active constants are given in the following table. They are recorded here to show how helpful the study of radio-activity has been in working out the composition of matter, and to give some idea of the magnitude of the numbers and the minuteness of the quantities dealt with.

Electric charge carried by each H atom in electrolysis 4.65 ?10^{-10} e.s.[1] Electric charge carried by each [alpha] particle 9.3 ?10^{-10} e.s. Number of atoms in 1 gram of H 6.2 ?10^{23} Mass of 1 atom of H 1.6 ?10^{-24} gram Number of molecules per cc. of any gas at standard pressure and temperature 2.72 ?10^{19} Number of [alpha] particles expelled per second per gram of radium itself 3.6 ?10^{10} Number of [alpha] particles expelled per second per gram of radium in equilibrium with its products 14.3 ?10^{10}

[1] The expression 10^{-10} means multiplying by .000,000,000,1; 10^{10} means multiplying by 10,000,000,000.

CHAPTER V

THE STRUCTURE OF THE ATOM

Properties of Radium

A study of the properties of radium will aid in throwing light upon the question as to the building up of the atom. First to be considered are the usual properties which distinguish an elementary body. Metallic radium has been prepared by a method similar to that used in the preparation of barium. It is a pure white metal, melting at 700? and far more volatile than barium. It rapidly alters on exposure to the air, probably forming a nitride. It energetically decomposes water and the product dissolves in the water. Its atomic weight is 226.

Radium forms a series of salts analogous in appearance and chemical action to those of barium. In the course of time they become colored, especially if mixed barium salts. The radiations from radium produce marked chemical effects in a number of substances. Carbon dioxide is changed into carbon, oxygen, and carbon monoxide, and the latter is changed into carbon and oxygen. Ammonia is dissociated into nitrogen and hydrogen; hydrochloric acid into chlorine and hydrogen. Oxygen is condensed into ozone. In general, the action upon gases appears to be similar to that of the silent electric discharge. Water is decomposed into hydrogen and oxygen. If moist radium chloride or a salt of radium containing water of crystallization is sealed in a glass tube, the gradual accumulation of hydrogen and oxygen will burst the tube.

The radiations rapidly decompose organic matter with the evolution of gases. Thus grease from stopcocks of apparatus used with radium or paraffin will give off carbon dioxide. Under an intense alpha radiation paraffin or vaseline become hard and infusible. White phosphorus is changed into red.

The action upon living tissue is most noteworthy, as its possible use as a remedial agent is dependent upon this. A small amount of a radium salt enclosed in a glass tube will cause a serious burn on flesh exposed to it. It therefore has to be handled with care and undue exposure to the radiations must be avoided. Cancer sacs shrivel up and practically disappear under its action. Whether the destruction of whatever causes the cancer is complete is at least open to serious doubt.

The coagulating effect upon globulin is interesting. When two solutions of globulin from ox serum are taken and acetic acid added to one while ammonia is added to the other, the opalescence in drops of the former is rapidly diminished on exposure to radium, showing a more complete solution, whereas the latter solution rapidly turns to a jelly and becomes opaque, indicating a greatly decreased solubility.

Energy Evolved by Radium

The greater part of the tremendous energy evolved by radium is due to the emission of the alpha particles, and in comparison the beta and gamma rays together supply only a small fraction. This energy may be measured as heat. It was first observed that a radium compound maintained a temperature several degrees higher than that of the air around it. The rate of heat production was later measured by means of an ice calorimeter and also by noting the strength of the current required to raise a comparison tube of barium salt to the same temperature. Both methods showed that the heat produced was at the rate of about 135 gram calories per hour. As the emission is continuous, one gram of radium would therefore emit about 1,180,000 gram calories in the course of a year. At the end of 2000 years it would still emit 590,000 gram calories per year. Such a production of energy so far surpasses all experience that it becomes almost inconceivable. It is futile to speak of it in terms of the heat evolved by the combustion of hydrogen, which is the greatest that can be produced by chemical means.

This effect is unaltered at low temperatures, as has been tested by immersing a tube containing radium in liquid air. It should be stated that these measurements were made after the radium had reached an equilibrium with its products; that is, after waiting at least a month after its preparation. The evolution of heat from radium and the radio-active substances is, in a sense, a secondary effect, as it measures the radiant energy transformed into heat energy by the active matter itself and whatever surrounds it. Let us repeat, therefore, that the total amount of energy pent up in a single atom of radium almost passes our powers of conception.

Necessity for a Disintegration Theory

The facts gathered so far justify and necessitate a theory which shall satisfactorily explain them, and since these phenomena are not caused by nor subject to the influence of external agencies, they must refer to changes taking place within the atom--in other words, a theory of disintegration. In the main, these facts may be summed up as the emission of certain radiations from known elemental matter: the material alpha particles with positive charge, the beta particles or negative electrons, and the gamma rays analogous to X rays. The emission of these rays results in the production of great heat. Then there is the law of transformations by which whole series of new elements are generated from the original element and maintain a constant equilibrium of growth and decay in the series. Lastly, we have the production of helium from the alpha particles.

Disintegration Theory

In explanation of these phenomena, Rutherford offered the hypothesis that the atoms of certain elements were unstable and subject to disintegration. The only elements definitely known to come under this description are the two having atoms of the greatest known mass, thorium (232) and uranium (238).

The atoms of uranium, for instance, are supposed to be not permanent but unstable systems. According to the hypothesis, about 1 atom in every 10^{18} becomes unstable each second and breaks up with a violent explosion for so small a mass of matter. One, or possibly two alpha particles are expelled with great velocity. This alpha particle corresponds to an atom of helium with an atomic weight of 4, and its loss reduces the original atomic weight to 234 with the formation of a new element, having changed properties corresponding to the new atomic weight. This new element is uranium X_{1}.

These new atoms are far more unstable than those of uranium, and the decomposition proceeds at a new rate of 1 in 10^{7} per second. So at a definite, measurable rate this stepwise disintegration proceeds. The explosions are not in all cases equally violent in going from element to element, nor are the results the same. Sometimes alpha particles alone are expelled, sometimes beta, or two of them together, as alpha and beta.

The new product may remain with the unchanged part of the original matter. Thus there would be an accumulation of it until its own decay balances its production, resulting eventually in a state of equilibrium.

Constitution of the Atom

In order to explain the electrical and optical properties of matter, the hypothesis was made that the atom consisted of positively and negatively electrified particles. Later it was shown that negative electrons exist in all kinds of matter. Various attempts were made to work out a model of such an atom in which these particles were held in equilibrium by electrical forces. The atom of Lord Kelvin consisted of a uniform sphere of positive electrification throughout which a number of negative electrons were distributed, and J. J. Thomson has determined the properties of this type as to the number of particles, their arrangement and stability.

Rutherford's Atom

According to Rutherford, the atom of uranium may be looked upon as consisting of a central charge of positive electricity surrounded by a number of concentric rings of negative electrons in rapid motion. The positively charged centre is made up of a complicated system in movement, consisting in part of charged helium and hydrogen atoms, and practically the whole charge and mass of the atom is concentrated at the centre. The central system of the atom is from some unknown cause unstable, and one of the helium atoms escapes from the central mass as an alpha particle.

There are, confessedly, difficulties connected with this conception of the atom which need not, however, be discussed here. Much remains to be learned as to the mechanics of the atom, and the hypothesis outlined above will probably have to be materially altered as knowledge grows. Perhaps it may have to be entirely abandoned in favor of some more satisfactory solution. Until such time it at least suffices as a mental picture around which the known facts group themselves. In this picture energy and matter lose their old-time distinctness of definition. Discrete subdivisions of energy are recognized which may be called charged particles without losing their significance. Some of these subdivisions charged in a certain way or with neutralized charge exhibit the properties of so-called matter.

Scattering of Alpha Particles

This conception of the atom would doubtless fail of much support were it not for certain experimental facts which lend great weight to it. Certain suppositions can be based on this theory mathematically reasoned out and tested by experiment. Predictions thus based on mathematical reasoning and afterward confirmed by experiment give a very convincing impression that truth lies at the bottom.

The first of these experimental proofs comes under the head of what is known as the scattering of the alpha particles, a phenomenon which, when first observed, proved hard to explain. If an alpha particle in its escape from the parent atom should come within the influence of the supposed outer electrical field of some other atom, it should be deflected from its course and, the intensity of the two charges being known, the angle of deflection could be calculated. For instance, if it came to what might be called a head-on collision with the positive central nucleus of another atom, it would recoil if it were itself of lesser mass, or would propel the other forward if that were the lighter.

The experiment is carried out by placing a thin metal foil over a radio-active body, as radium C, which expels alpha particles with a high velocity, and counting the number of alpha particles which are scattered through an angle greater than 90?and so recoil toward their source. This has been done by a number of investigators and it has been found that the angle of scattering and the number of recoil particles depend upon the atomic weight of the metal used as foil. For example, if gold is used, the number of recoil atoms is one in something less than 8,000.

Taking the atomic weight of gold into consideration, Rutherford calculated mathematically that this was about the number which should be driven backward. But he went further and calculated also the number which should be returned by aluminum, which has an atomic weight of only about one-seventh that of gold. Two investigators determined experimentally the number for aluminum and their results agreed with Rutherford's calculations.

The metals from aluminum to gold have been examined in this way. The number of recoil particles increases with the atomic weight of the metal. Comparing experiment with theory, the central charge in an atom corresponds to about one-half the atomic weight multiplied by the charge on an electron, or, as it is expressed, 1/2 Ae.

There is only one lighter atom than helium, namely, hydrogen, which has a mass only one-fourth as great. When alpha particles are discharged into hydrogen, a few of the latter atoms are found to be propelled to a distance four times as great as that reached by the alpha particles.

Stopping Power of Substances

Parallel with the experiments mentioned, there is what is called the stopping power of substances. This means the depth or thickness of a substance necessary to put a stop to the course of the alpha particles. This gives the range of the alpha particles in such substances and is connected in a simple way with the atomic weight, that is, it is again fixed by the mass of the opposing atom. This stopping power of an atom for an alpha particle is approximately proportional to the square root of its atomic weight.

Considering gases, for instance, if the range in hydrogen be 1, then the range in oxygen, the atomic weight of which is 16, is only $(1/16)^{1/2}$ or 1/4. Generally in the case of metals the weight of matter per unit area required to stop the alpha particle is found to vary according to the square root of the atomic weight of the metal taken.

CHAPTER VI

RADIO-ACTIVITY AND CHEMICAL THEORY

Influence upon Chemical Theory

It can easily be seen that the revelations of radio-activity must have a far-reaching effect upon chemical theory, throwing light upon, and so bringing nearer, the solution of some of the problems which have been long discussed without arriving at any satisfactory solution. The so-called electro-chemical nature of the elements will certainly be made much clearer. The changes in valence should become intelligible and valence itself should be explained. A fuller understanding of the ionization of electrolytes also becomes possible. As these matters are debatable and the details are still unsettled, it is scarcely appropriate to give here the hypotheses in detail or to enter into any discussion of them. But the promise of solution in accord with the facts is encouraging.

The Periodic System

Such progress has been made, however, in regard to a better understanding of the Periodic System that the new facts and their interpretation may well be given. No reliable clue to the meaning of this system and the true relationship between the elements had been found up to the time when new light was thrown upon it by the discoveries of radio-activity. The underlying principle was unknown and even the statement of what was sometimes erroneously called the Periodic Law was manifestly incorrect and its terms were ignored.

Basis of the Periodic System

The ordinary statement of the fundamental principle of the Periodic System has been that the properties of the elements were periodic functions of the atomic weights, and that when the elements were arranged in the order of their atomic weights they fell into a natural series, taking their places in the proper related groups.

In accepting this, the interpretation of function was both unmathematical and vague, and the order of the atomic weights was not strictly adhered to but unhesitatingly abandoned to force the group relationship. Wherever consideration of the atomic weight would have placed an element out of the grouping with other elements to which it was clearly related in physical and chemical properties, the guidance of these properties was accepted and that of the atomic weights disregarded. Such shiftings are noted in the cases of tellurium and iodine; cobalt and nickel; argon and potassium. It was most helpful that, following the order of atomic weights, the majority of the elements fell naturally into their places. Otherwise the generalization known as the Periodic System might have remained for a long time undiscovered and the progress of chemistry would have been greatly retarded.

Influence of Positive Nucleus

It is evident that the order of the elements is determined by something else than their atomic weights. From the known facts of radio-activity it would seem that this determining factor is the positive nucleus. And this nucleus also determines the mass or weight of the atom. Taking the elements in their order in the Periodic Series and numbering the positions held by them in this series as 1, 2, 3, etc., we get the position number or what is called the atomic number. This designates the order or position of the element in the series. We must learn that this number marks a position rather than a single element, a statement which will be explained later.

Determination of the Atomic Number

Since the atomic weight is unreliable as a means of settling the position of an element in the series and so fixing its atomic number, how is this number to be determined? Of course, one answer to this question is that we may rely upon a consideration of the general properties, as has been done in the past. Fortunately, other methods have been found by which this may be confirmed. For instance, the stopping and scattering power of the element for alpha particles has been suggested and successfully used.

Use of X-Ray Spectra

A most interesting method is due to Moseley's observations upon the X-ray spectra of the various elements. It has been found that crystals, such as those of quartz, have the power of reflecting and defining the X rays. The spectra given by these rays can be photographed and the wave lengths measured. These X rays are emitted by various substances under bombardment by the cathode rays (negative electrons) and have great intensity and very minute wave lengths. Moseley made use of various metals as anti-cathodes for the production of these rays. These metals ranged from calcium to zinc in the Periodic System. In each case he observed that two characteristic types of X rays of definite intensity and different wave lengths were emitted. From the frequency of these waves there is deduced a simple relation connected with a fundamental quantity which increases in units from one element to the next. This is due to the charge of the positive central nucleus. The number found in this way is one less than the atomic number. Thus the number for calcium is 19 instead of 20 and that for zinc is 29 instead of 30. So, by adding 1 to the number found the atomic number is obtained.

The atomic weight can usually be followed in fixing the atomic number, but where doubt exists the method just given can be resorted to. Thus doubt arises in the case of iron and nickel and cobalt. This would be the order according to the atomic weights. The X-ray method gives the order as iron, cobalt, and nickel, and this is the accepted order in the Periodic System.

Changes Caused by Ray Emission

On studying the properties of the elements in a transformation series in connection with the ray emission which produced them, it was seen that these properties were determined in each case by the nature of the ray emitted from the preceding transformation product or parent element.

Atomic Weight Losses

Each alpha particle emitted means a loss of 4 in the atomic weight. This is the mass of a helium atom. Thus from uranium with an atomic weight of 238 to radium there is a loss of three alpha particles. Therefore, 12 must be subtracted from 238, leaving 226, which agrees closely with the atomic weight of radium as actually determined by the ordinary methods. Uranium X{1}, then, would have an atomic weight of 234 and that of ionium would be 230. The other intermediate elements, whose formation is due to the loss of beta particles only, show no decrease in atomic weight.

Lead the End Product

From uranium to lead there is a loss of 8 alpha particles, or 32 units in atomic weight. This would give for the final product an atomic weight of 206. The atomic weight of lead is 207.17. It is not at all certain that the final product of this series is ordinary lead. The facts are such that they would lead one to think that it is not. It is known only that the end product would probably be some element closely resembling lead chemically and hence difficult or impossible to separate from it. Several accurate determinations of lead coming from uranium minerals, which always carry this element and in an approximately definite ratio to the amount of uranium present, show atomic weights of 206.40; 206.36; and 206.54. Even the most rigid methods of purification fail to change these results. The lead in these minerals might therefore be considered as coming in the main from the disintegration of the uranium atom and, though chemically resembling lead, as being in reality a different element with different atomic weight.

Furthermore, in the thorium series 6 alpha particles are lost before reaching the end product, which again is perhaps the chemical analogue of lead. The atomic weight here should be 232 less 24, or 208. Determinations of the atomic weight of lead from thorite, a thorium mineral nearly free from uranium, gave 208.4.

The end product of the actinium series is also an element resembling lead, but both the beginning and ending of this series are still in obscurity.

Changes of Position in the Periodic System

The loss of 4 units in the atomic weight of an element on the expulsion of an alpha particle is accompanied by a change of chemical properties which removes the new element two groups toward the positive side in the Periodic System.

Thus ionium is so closely related to thorium and so resembles it chemically that it is properly classed along with thorium as a quadrivalent element in the fourth group. Ionium expels an alpha particle and becomes radium, which is a bivalent element resembling barium belonging to the second group. Radium then expels an alpha particle and becomes the gas, radium emanation, which is an analogue of argon and belongs to the zero group. Other instances might be cited which go to show that in all cases the loss of an alpha particle makes a change of two places toward the left or positive side of the System.

Changes from Loss of Beta Particles

The loss of a beta particle causes no change in the atomic weight but does cause a shift for each beta particle of one group toward the right or negative side of the System. Two such losses, then, will counterbalance the loss of an alpha particle and bring the new element back to the group originally occupied by its progenitor. Thus uranium in the sixth group loses an alpha particle and the product UX{1} falls in the fourth group. One beta particle is then lost and UX{2} belonging to the fifth group is formed. With the loss of one more beta particle the new element returns to the sixth group from which the transformation began.

The table on page 48, as adapted from Soddy, affords a general view of these changes.

Isotopes

An examination of the table will show a number of different elements falling in the same position in a group of the Periodic System irrespective of their atomic weights. These are chemically inseparable so far as the present limitations of chemical analysis are concerned. Even the spectra of these elements seem to be identical so far as known. This identity extends to most of the physical properties, but this demands much further investigation. For this new phenomenon Soddy has suggested the word isotope for the element and isotopic for the property, and these names have come into general use.

Manifestly, we have come across a phenomenon here which quite eliminates the atomic weight as a determining factor as to position in the Periodic or Natural System or of the elemental properties in general. All of the properties of the bodies which we call elements, and consequently of their compounds and hence of matter in general, seem to depend upon the balance maintained between the charges of negative and positive electricity which, according to Rutherford's theory, go to make up the atom.

It is evident that any study of chemical phenomena and chemical theory is quite incomplete without a study of radio-activity and the transformations which it produces.

Radio-activity in Nature

In concluding this outline of the main facts of radio-activity, it is of interest to discuss briefly the presence of radio-active material on this planet and in the stars. Facts enough have been gathered to show the probable universality of this phenomenon of radio-activity. Whether this means solely the disintegration of the uranium and thorium atoms, or whether other elements are also transformed under the intensity of the agencies at work in the universe, is of course a question as yet unsolved.

Radio-active Products in the Earth's Crust

The presence of uranium and thorium widely distributed throughout the crust of the earth would lead to the conclusion that their disintegration products would be found there also. Various rocks of igneous origin have been examined revealing from 4.78 $?10^{-12}$ to 0.31 $?10^{-12}$ grams of radium per gram of the rock. Aqueous rocks have shown a lesser amount, ranging from 2.92 $?10^{-12}$ to 0.86 $?10^{-12}$ grams. As the soil is formed by the decomposition of these rocks, radium is present in varying amounts in all kinds of soil.

Presence in Air and Soil Waters

As radium is transformed into the gaseous emanation, this will escape wherever the soil is not enclosed. For instance, a larger amount of radio-activity is found in the soil of caves and cellars than in open soils. If an iron pipe is sunk into a soil and the air of the soil sucked up into a large electroscope, the latter instrument will show the effect of the rays emitted and will measure the degree of activity. Also the interior of the pipe will receive a deposit of the radio-active material and will show appreciable radio-activity after being removed from the soil.

This radium emanation is dissolved in the soil waters, wells, springs, and rivers, rendering them more or less radio-active, and sometimes the muddy deposit at the bottom of a spring shows decided radio-activity.

The emanation also escapes into the air so that many observations made in various places show that the radium emanation is everywhere present in the atmosphere. Neither summer nor winter seems to affect this emanation, and it extends certainly to a height of two or three miles. Rain, falling through the air, dissolves some of the emanation, so that it may be found in freshly-fallen rain water and also in freshly-fallen snow. Radio-active deposits are found upon electrically charged wires exposed near the earth's surface.

As helium is the resulting product of the alpha particles emitted by the emanation and other radio-active bodies, it is found in the soil air, soil waters, and atmosphere.

Average measurements of the radio-activity of the atmosphere have led to the calculation that about one gram of radium per square kilometer of the earth's surface is requisite to keep up the supply of the emanation.

A number of estimates have been given as to the heat produced by the radio-active transformations going on in the material of this planet. Actual data are scarce and mere assumptions unsatisfactory, so little that is worth while can be deduced. It is possible that this source of heat may have an appreciable effect upon or serve to balance the earth's rate of cooling.

Cosmical Radio-activity

Meteorites of iron coming from other celestial bodies have not shown the presence of radium. Aerolites or stone meteorites have been found to contain as much as similar terrestrial rock. Since the sun contains helium and some stars show its presence as predominating, this suggests the presence of radio-active matter in these bodies. In addition, the spectral lines of uranium, radium, and the radium emanation have been reported as being found in the sun's spectrum and also in the new star, Nova Geminorum 2. These observations await further investigation and confirmation. So far as the sun's chromosphere is concerned, the possible amount of radium present would seem to be very small. If this is true, radio-active processes could have little to do with the sun's heat. The statement is made by Rutherford that indirect evidence obtained from the study of the aurora suggests that the sun emits rays similar in type to the alpha and beta rays. Such rays would be absorbed, and the gamma rays likewise, in passing through the earth's atmosphere and so escape ordinary observation. All of this is but further evidence of the unity of matter and of forces in the universe.

###

www.ingramcontent.com/pod-product-compliance
Lightning Source LLC
Chambersburg PA
CBHW070923180526
45168CB00005B/2133